Looking for Shapes

by Margie Burton, Cathy French, and Tammy Jones

Table of Contents

If you go walking around outside,
you will see many shapes. You
can see shapes on the houses
and in trees and plants.
You can find shapes everywhere.

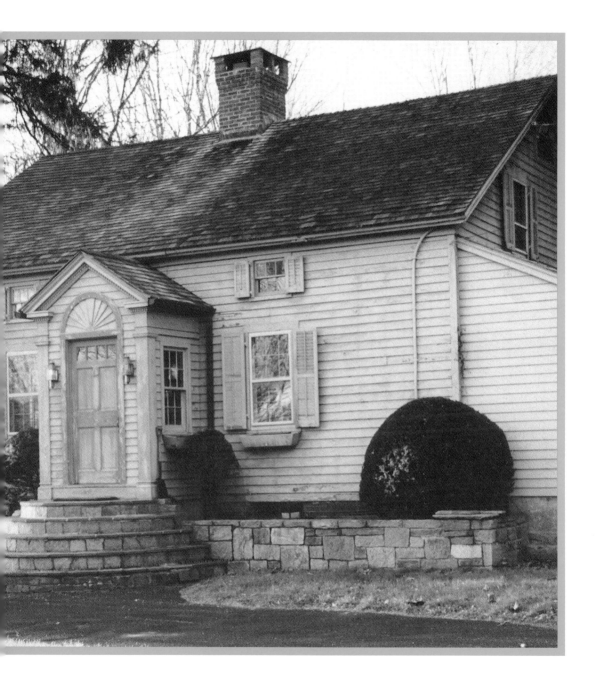

What Is a Rectangle?

A rectangle is a shape that has
four sides and four square corners.
Some have two long sides
and two short sides. Sometimes
the sides are the same length.

A rectangle has four sides and four square corners.

You might see some rectangles when you go walking.

This wall is made up of rectangles. Can you find the long sides?

The door on this house is a rectangle.
The shutters are rectangles, too.

What Is a Square?

A square is a kind of rectangle.
A square is a shape that has four sides
that are the same length.

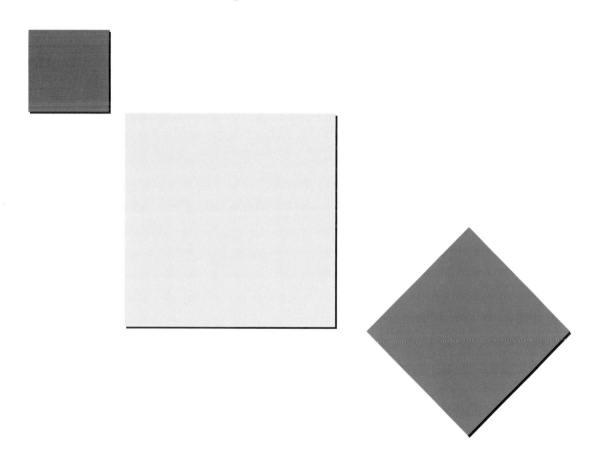

A square has four equal sides and four square corners.

If you walk into town, you might see
some square windows.

The town clock is on a square.

What Is a Triangle?

A triangle is a shape that has
three sides. The sides come
together to make points on
each end.

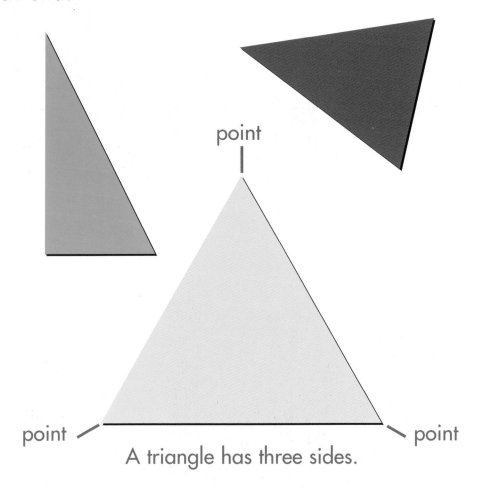

A triangle has three sides.

If you keep walking, you might see some triangles.

Do you see the triangle? Can you find the sides?

This tree is shaped like a triangle, too.

Here is a sailboat that has a sail in the shape of a triangle.

What Is a Circle?

A circle is a shape that is round.
Some circles are bigger than others.

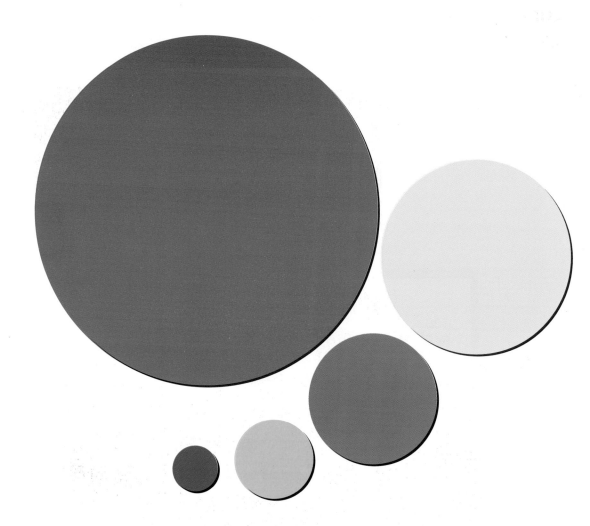

You can see many circles
around town.

Look at the wheels on trucks,
cars, and buses. They are
all circles.

Some signs are circles.

You can find animals that
have some circles, too.
Can you find some spots
that look like a circle?

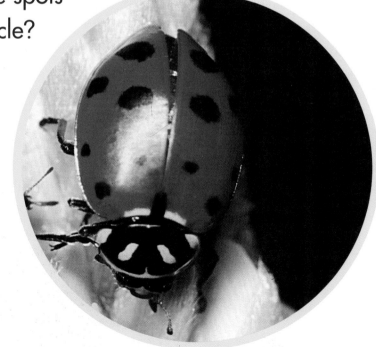

Can you find the rectangles?

Can you find the squares?

Can you find the triangles?

Can you find the circles?

You can find shapes everywhere.